Skill Builders
Time and Money

by Clareen Arnold

Welcome to Rainbow Bridge Publishing's Skill Builders series. The Skill Builders series is designed to make learning fun and rewarding.

This workbook is based on national standards and is designed to reinforce classroom math skills and strategies for students learning to do math with time and money. This workbook holds students' interest with the right mix of challenge, imagination, and instruction. The diverse assignments teach about time and money while giving the students something to think about—from space shuttles to school lunch.

A critical thinking section includes exercises to help develop higher-order thinking skills.

Learning is more effective when approached with an element of enthusiasm. That's why the Skill Builders combine entertaining and academically sound exercises with appealing graphics and engaging themes—to make reviewing basic skills at school or home fun and effective, for both you and your budding scholars.

© 2004 RBP Books
All rights reserved.
www.summerbridgeactivities.com

Table of Contents

The Penny
  The Penny 3
  Find the Penny 4
  Counting Pennies 5
  Adding Up Your Pennies 6
  Buying Toys 7
  Buying Stuffed Animals 9
  Adding Money 10
  Subtracting Money 11
  Equations with Money 12
  Probability 13

The Nickel
  The Nickel 14
  Counting with Nickels 15
  Counting Nickels and
    Pennies 18
  Matching Coins to
    Price Tags 20
  Buying Toys 21
  Buying Snacks 22
  Buying Lunch 23
  Buying Fruit 24

The Dime
  The Dime 25
  Counting Dimes 26
  How to Make Ten Cents 27
  Counting Dimes 28
  Dimes and Pennies 29
  Buying Toys 30
  Counting Money 31
  Getting Change 32
  Subtracting Money 33

The Quarter
  The Quarter 34
  How to Make
    Twenty-Five Cents 35
  Adding Up to a Quarter 36
  Spending Money 37
  Bank a Quarter 38
  Making Change 39
  Money Maze 40

Critical Thinking Skills
  Spending Money 41

Time
  Clocks 44
  Numbering a Clock 45
  Minute Hand and
    Hour Hand 46
  Time by the Hour 47
  Writing in the Hour 49
  Drawing the Hour Hand
    on the Clock 50
  Writing the Time 51
  Digital Clocks 52
  Putting the Hands
    on the Clocks 54
  Matching Time
    on the Clock 55
  Telling Time
    One Hour Ahead 57
    One Hour Behind 58
  Read a Story
    about Time 59
  Writing the Minutes 61
  Telling Time to the
    Half Hour 62
  Writing Time to the
    Half Hour 64
  Matching Time to the
    Half Hour 65
  Putting Hands
    on the Clock 66
  Matching Time to the
    Half Hour 68
  Digital Clocks to the
    Half Hour 69
  Matching Digital Clocks
    to Face Clocks 70

Critical Thinking Skills
  Counting Hours 71
  Counting Hours and
    Minutes 73

Answer Key 75

The Penny

This is a penny.
1 penny = 1 cent
1 penny = 1¢

It has 2 sides.

front   back

This is the cent symbol: ¢
Now you make the cent symbol. _____

Color the correct number of pennies.

1. Color 5 pennies = 5¢

2. Color 3 pennies = 3¢

3. Color 7 pennies = 7¢

4. Color 1 penny = 1¢

Find the Penny

Find each penny. Color it brown.

Good work! How many pennies did you find? _____

Counting Pennies

Count each group of pennies and write the amount.

1.

3 pennies = 3¢

2.

___ pennies = ___¢

3.

___ penny = ___¢

4.

___ pennies = ___¢

5.

___ pennies = ___¢

6.

___ pennies = ___¢

Adding Up Your Pennies

You have been saving your pennies. Count to see how many pennies you have.

Week one:
1¢ + 1¢ = 2¢

Week two:
____¢ + ____¢ + ____¢ = ____¢

Week three:
____¢ + ____¢ + ____¢ + ____¢ = ____¢

Week four:
____¢ + ____¢ + ____¢ + ____¢ + ____¢ = ____¢

Total pennies saved: _____.

Buying Toys

How much does each toy cost? Color the correct number of pennies and write the amount.

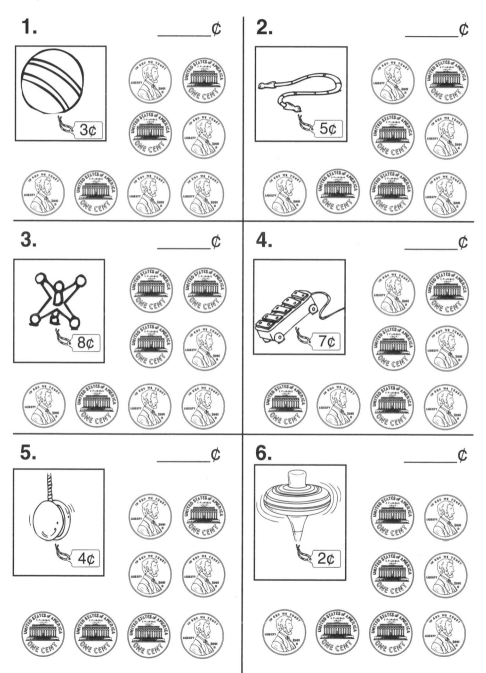

Buying Toys

Color the pennies you need.

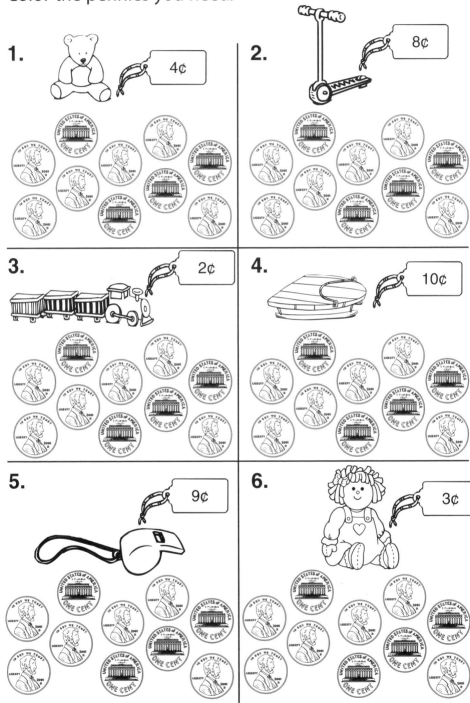

Buying Stuffed Animals

Write how many cents each stuffed animal costs.

1.

2.

_7_ ¢

____ ¢

3.

4.

____ ¢

____ ¢

Adding Money

Find out how much money there is by adding up all of the pennies.

1.    4¢
              + 2¢
              ___
              6¢

2.   1¢
              + 3¢
              ___

3.   5¢
              + 5¢
              ___

4.   3¢
              + 2¢
              ___

5.   4¢
       + 3¢
       ___

6.   4¢
       + 1¢
       ___

Subtracting Money

Find how much money is left. Cross out pennies and then subtract.

1. 4¢
  − 2¢
  2¢

2. 3¢
  − 1¢

3. 5¢
  − 5¢

   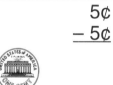

4. 6¢
  − 2¢

5. 4¢
  − 3¢

6. 8¢
  − 1¢

Equations with Money

The parts are the same. The sums are the same.

1¢ + 4¢ = 5¢   4¢ + 1¢ = 5¢

Find out how many cents there are in all. Use real 's to help.

A. 2¢ + 3¢ = _____ ¢   **B.** 3¢ + 2¢ = _____ ¢

C. 6¢ + 4¢ = _____ ¢   **D.** 4¢ + 6¢ = _____ ¢

E. 1¢ + 5¢ = _____ ¢   **F.** 5¢ + 1¢ = _____ ¢

G. 2¢ + 4¢ = _____ ¢   **H.** 4¢ + 2¢ = _____ ¢

I. 7¢ + 3¢ = _____ ¢   **J.** 3¢ + 7¢ = _____ ¢

Probability

Allie wants to see what the chance is of a penny coming up heads. Toss a penny in the air. When it lands, look to see if it landed with heads (the front) or tails (the back) facing up. Make a mark in the correct column after each toss. Toss the penny 8 times.

Heads	Tails
1. _____	_____
2. _____	_____
3. _____	_____
4. _____	_____
5. _____	_____
6. _____	_____
7. _____	_____
8. _____	_____

Number of heads = _____   Number of tails = _____

The Nickel

The Penny

This is a penny.
1 penny = 1 cent
1 penny = 1¢
Color it brown.

It has 2 sides.

front   back

The Nickel

This is a nickel.
1 nickel = 5 cents
1 nickel = 5¢
Color it silver.

It has 2 sides.

front   back

 + + + + =

You count cents with nickels by 5's. Count by 5's. Fill in the blanks.

5	10	15	20	___
30	___	40	45	___
55	60	___	___	75
___	85	___	95	___

Counting with Nickels

Count by 5's. Write how much the nickels are worth on the line.

1.

This is _____ cents.

2.

This is _____ cents.

3.

This is _____ cents.

4.

This is _____ cents.

5.

This is _____ cents.

6.

This is _____ cents.

Counting Nickels

Color the correct number of nickels.

1.

Color 10 cents.

2.

Color 5 cents.

3.

Color 20 cents.

4.

Color 25 cents.

5.

Color 30 cents.

6.

Color 15 cents.

Counting Nickels

Count each group of nickels and write the value on the line.

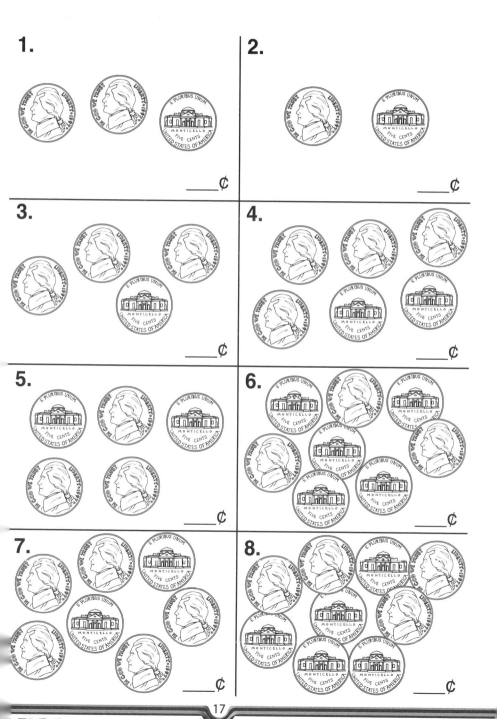

Counting Nickels and Pennies

You have been saving your money. Count to see how much money you have saved.

Week one:
5¢ + 1¢ = 6¢

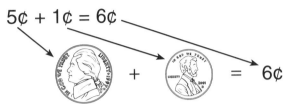

Week two:
____¢ + ____¢ + ____¢ = ____¢

Week three:
____¢ + ____¢ + ____¢ + ____¢ = ____¢

Week four:
____¢ + ____¢ + ____¢ + ____¢ + ____¢ = ____¢

Count all of the money. What is the total amount of money you have saved in four weeks? _____

Counting Nickels and Pennies

Write how much money you have in nickels and pennies.

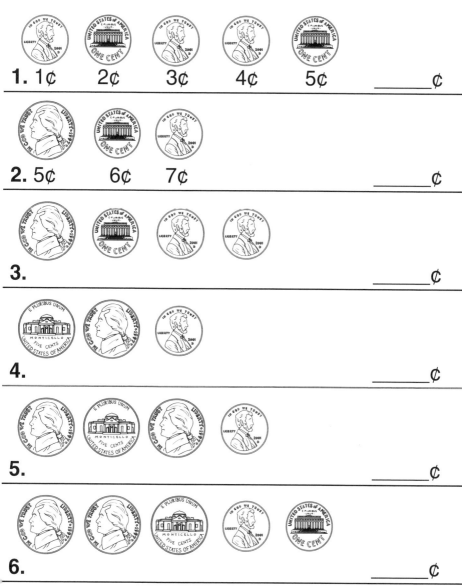

1. 1¢ 2¢ 3¢ 4¢ 5¢ _____ ¢

2. 5¢ 6¢ 7¢ _____ ¢

3. _____ ¢

4. _____ ¢

5. _____ ¢

6. _____ ¢

7.   _____ ¢

Matching Coins to Price Tags

Match the coin groups with the price tags.

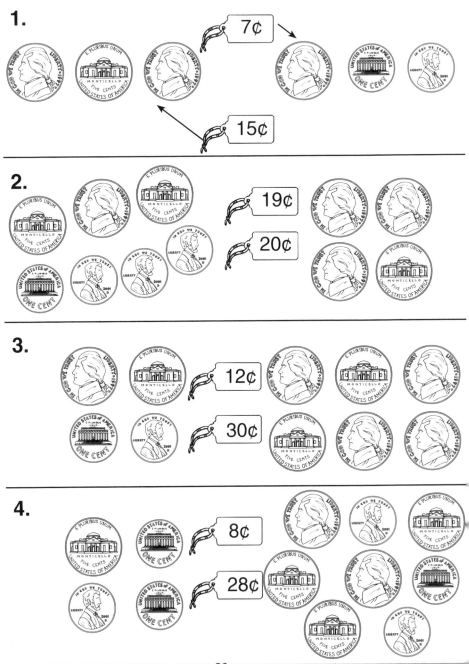

Buying Toys

Color the coins you need to buy each item.

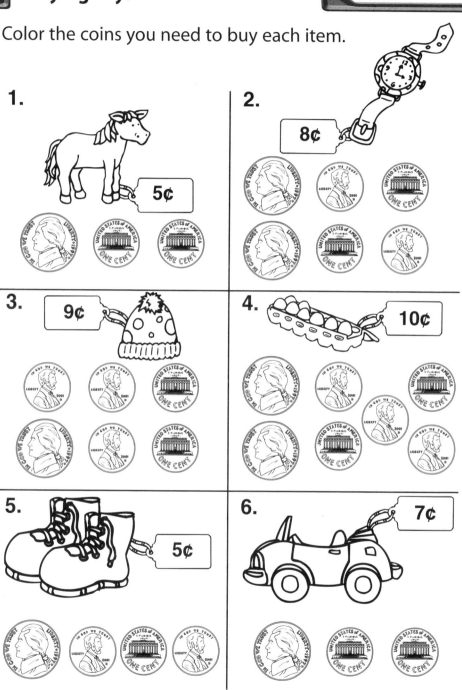

Count all of the colored-in coins. How much money have you spent? _____

Buying Snacks

It is your turn to buy snacks. How much does each snack cost? Count the coins. Write the value on the bag.

Buying Lunch

Each day you get money for school lunch. Count your money and write how much you have. Then, circle the food choice that matches the money you have.

You have: That's _____ ¢

Pick your drink!   20¢   10¢   15¢

You have: That's _____ ¢

   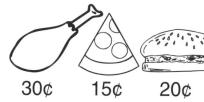

Pick your main course!   30¢   15¢   20¢

You have: That's _____ ¢

   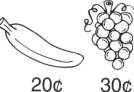

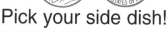

Pick your side dish!   20¢   30¢   35¢

You have: That's _____ ¢

Pick your dessert!   10¢   20¢   15¢

Buying Fruit

You stopped at a fruit stand to buy fruit.
Use the fruit you bought to write an addition sentence for each problem.

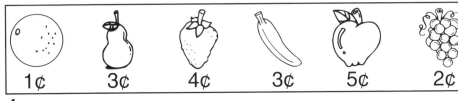

1.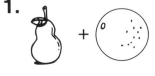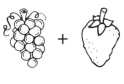

   3¢ + 1¢ = 4¢ _____ _____

2.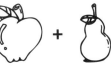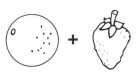

   _____ _____ _____

3.

   _____ _____ _____

4.

   _____ _____ _____

The Dime

The Penny

This is a penny.
1 penny = 1 cent
1 penny = 1¢
Color it brown.

It has 2 sides.

front   back

The Nickel

This is a nickel.
1 nickel = 5 cents
1 nickel = 5¢
Color it silver.

It has 2 sides.

front   back

The Dime

This is a dime.
1 dime = 10 cents
1 dime = 10¢
Color it silver.

It has 2 sides.

front   back

You count cents with dimes by 10's. Count by 10's. Fill in the blanks.

10   20   ___   40   ___

60   ___   80   90   ___

Counting Dimes

A dime is worth 10 cents, so we count dimes with 10's. Count the dimes in each box. Then, write the total value on each line.

1.

This is _____ cents.

2.

This is _____ cents.

3.

This is _____ cents.

4.

This is _____ cents.

5.

This is _____ cents.

6.

This is _____ cents.

7.

This is _____ cents.

8.

This is _____ cents.

How to Make Ten Cents

These are the ways you can make 10 cents.

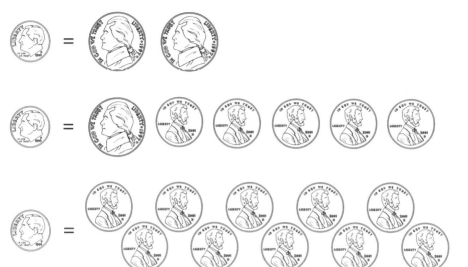

Now, draw the three different ways to make ten cents.

🪙 =

🪙 =

🪙 =

Counting Dimes

Match each set of dimes to the correct amount.

1. 10¢

2. 20¢

3. 30¢

4. 40¢

5. 50¢

6. 60¢

7. 70¢

8. 80¢

Dimes and Pennies

Count. Write how much money you have in dimes and pennies.

1.  __12__ ¢

2.  _____ ¢

3.   _____ ¢

4.    _____ ¢

5.  _____ ¢

6.   _____ ¢

7.   _____ ¢

Buying Toys

Color the coins you need to buy each item.

1. 15¢

2. 35¢

3. 20¢

4. 65¢

5. 40¢

6. 70¢

7. 95¢

Counting Money

How much money does each person have? Count the money. Then, circle the correct amount.

1. Brooke	53¢ 40¢
2. Jason	66¢ 79¢
3. Jeffrey	32¢ 29¢
4. Abbie	81¢ 92¢
5. Cory	54¢ 45¢
6. Kayce	96¢ 91¢

Getting Change

Your parents take you shopping. In the store window, you can see toys on the shelves. Read the problem. Write the subtraction sentence. Find out how much money each person has left.

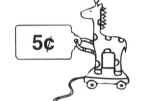

1. Michael has 8¢. He buys a giraffe.

____¢ – ____¢ = ___¢

2. Kristin has 10¢. She buys a top.

____¢ – ____¢ = ___¢

3. Kenny has 6¢. He buys a bear.

____¢ – ____¢ = ___¢

4. Kayla has 12¢. She buys a ball.

____¢ – ____¢ = ___¢

5. Jonathan has 18¢. He buys a kite.

____¢ – ____¢ = ___¢

6. Ruby has 20¢. She buys a jack.

____¢ – ____¢ = ___¢

Subtracting Money

Fill in the blank in the first column. In the middle column, draw the picture from below that makes the math problem correct.

I had ____ ¢.	I bought	I have 2¢ left.
I had ____ ¢.	I bought	I have 5¢ left.
I had ____ ¢.	I bought	I have 1¢ left.

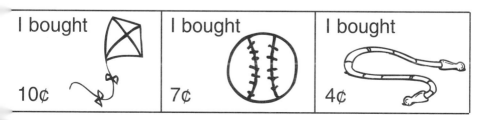

I bought 10¢	I bought 7¢	I bought 4¢

The Quarter

The Penny

This is a penny.
1 penny = 1 cent
1 penny = 1¢
Color it brown.

It has 2 sides.

front   back

The Nickel

This is a nickel.
1 nickel = 5 cents
1 nickel = 5¢
Color it silver.

It has 2 sides.

front   back

The Dime

This is a dime.
1 dime = 10 cents
1 dime = 10¢
Color it silver.

It has 2 sides.

front   back

The Quarter

This is a quarter.
1 quarter = 25 cents
1 quarter = 25¢
Color it silver.

It has 2 sides.

front   back

You count cents with quarters by 25's. Count by 25's. Fill in the blanks.

25   ____   75   100   ____   150   ____   200

How to Make Twenty-Five Cents

Look at some of the different ways you can make 25 cents.

25¢

10¢ + 10¢ + 5¢ = 25¢

10¢ + 5¢ + 5¢ + 5¢ = 25¢

5¢ + 5¢ + 5¢ + 5¢ + 5¢ = 25¢

10¢ + 5¢ + 5¢ + 1¢ + 1¢ + 1¢ + 1¢ + 1¢ = 25¢

Draw as many combinations of coins as you can to make 25¢.

Adding Up to a Quarter

Count each group of coins. Does it equal a quarter? Write yes or no.

1. __Yes__

2. ____

3. ____

4. ____

5. ____

6. ____

7. ____

8. ____

Spending Money

Count each set of coins. Then, look at the price tag. One more coin is needed to equal the amount on the price tag. Write the amount of the coin.

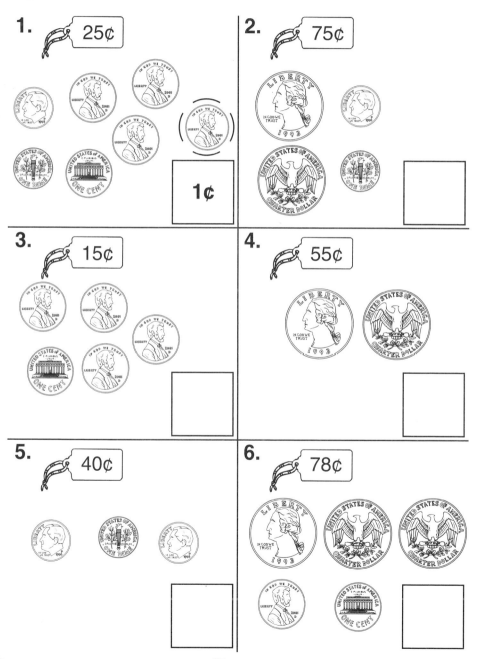

Bank a Quarter

One player is the banker. Take turns throwing a die. You get pennies from the banker for the number on the die. Trade your pennies for nickels and dimes. When you trade for a quarter, you win.

     trade →
Trade 5 pennies for a nickel.

  trade →
Trade 2 nickels for a dime.

   trade →
Trade 2 dimes and 1 nickel for a quarter.

You win!

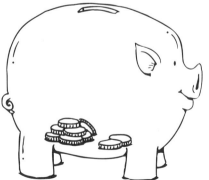

Making Change

How many ways can you make 25¢ using dimes, nickels, and pennies?

Dimes	0				
Nickels	4				
Pennies	5				

Think of ways to make 16¢.

Dimes	0				
Nickels	3				
Pennies	1				

Money Maze

How much money will the boy find on his way through the the maze? Money found on wrong turns should not be counted.

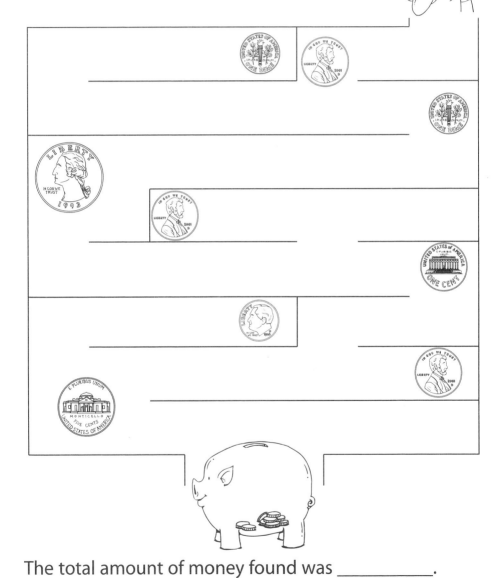

The total amount of money found was _____.

Spending Money

Melinda has 36¢. She wants to spend all of it. Circle the toys she can buy. Is there more than one way to spend all of the money?

Critical Thinking Skills

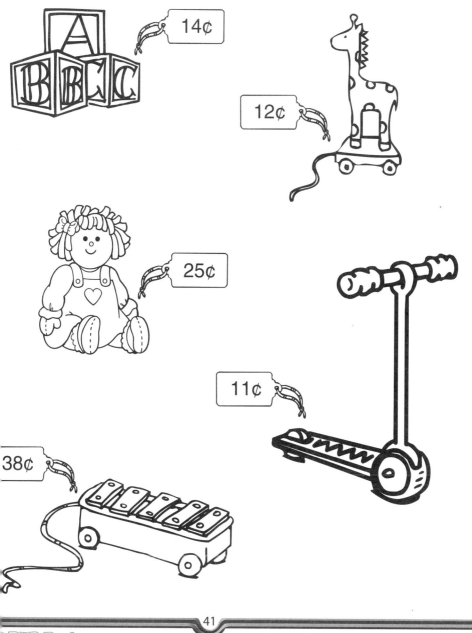

Spending Money

Show what coins you would use to buy each toy.

1. _____ quarters
   _____ dimes
   _____ nickels
   _____ pennies

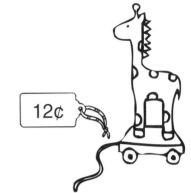
12¢

2. _____ quarters
   _____ dimes
   _____ nickels
   _____ pennies

26¢

3. _____ quarters
   _____ dimes
   _____ nickels
   _____ pennies

38¢

4. _____ quarters
   _____ dimes
   _____ nickels
   _____ pennies

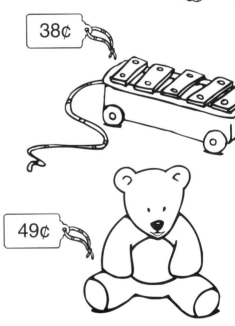
49¢

Spending Money

Answer these money questions.

1.

   How much money do you have? _____
   How much money does the dog cost? _____
   Do you need more money? _____
   How much more do you need? _____

2.

   How much money do you have? _____
   How much money does the cat cost? _____
   Do you need more money? _____
   How much more do you need? _____

3.

   How much money do you have? _____
   How much money does the fish cost? _____
   Do you need more money? _____
   How much more do you need? _____

Clocks

There are many clocks and watches in the world. Circle the ones you use at home.

Numbering a Clock

A clock has different parts to tell us the time.

A clock face has numbers on it. Write the numbers around the clock. Start with 1 and write to 12. These are the hours on a clock.

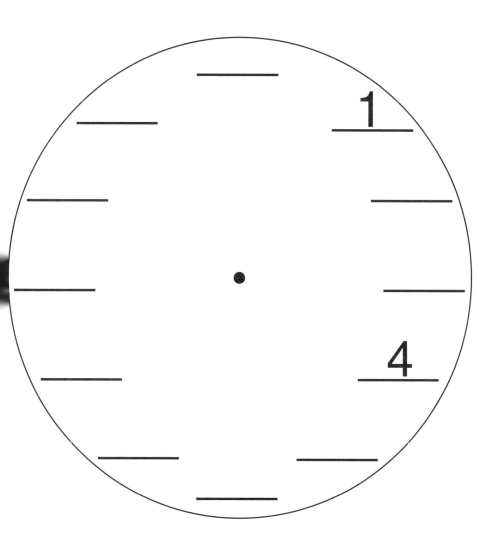

Minute Hand and Hour Hand

There are 2 hands on a clock.

The big hand is called the **minute hand**. Find the minute hand and color it blue.

The little hand is called the **hour hand**. Find the hour hand and color it orange.

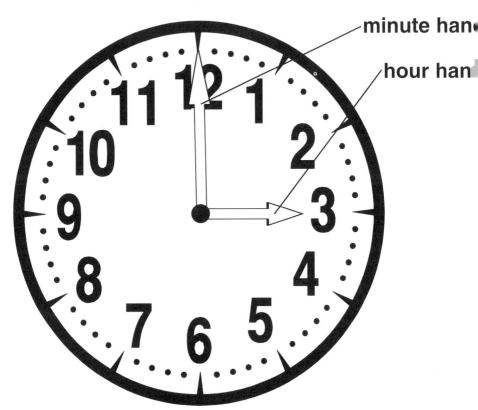

1. The minute hand is on the _____.

2. The hour hand is on the _____.

This clock tells us it is 3 o'clock.

Time by the Hour

When the minute hand (big hand) points to 12, it is time to start a new hour. The hour hand (small hand) points to the name of the hour.

Look at the clock. Trace the little hand to the 5. The minute hand is on _____. The hour hand is on _____. It is _____ o'clock.

Now look at the clocks below. Write what time it is.

1.
It is _____ o'clock.

2.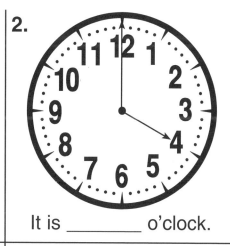
It is _____ o'clock.

3.
It is _____ o'clock.

4.
It is _____ o'clock.

Time by the Hour

Circle the correct hour.

1.
   1 2 3

2.
   3 4 5

3.
   5 6 7

4.
   7 8 9

5.
   9 10 11

6.
   11 12 1

Writing in the Hour

Look at each clock. What number does the hour hand point to? Write the number on the clock and in the box.

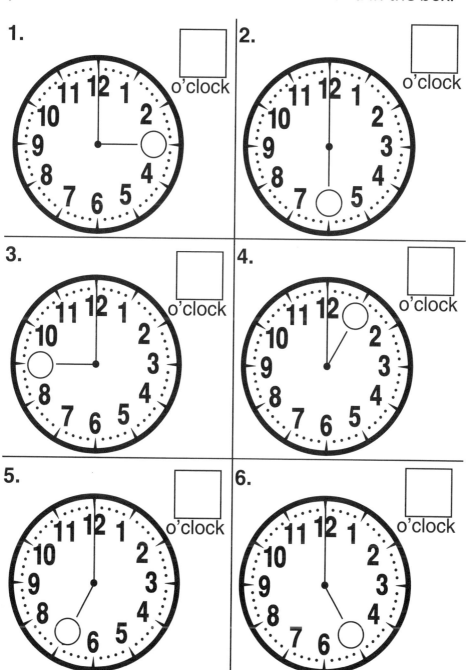

Drawing the Hour Hand on the Clock

Draw the hour hand on each clock.

1.
8 o'clock

2.
6 o'clock

3.
3 o'clock

4.
9 o'clock

5.
2 o'clock

6.
11 o'clock

Writing the Time

What time is it?
Look at the time on the clock. Then, write the time.

1.

____8____ o'clock

2.

_____ o'clock

3.

_____ o'clock

4.

_____ o'clock

5.

_____ o'clock

6.

_____ o'clock

Digital Clocks

This is a digital clock. It tells time with numbers. First, it tells the hour, then the minutes.

$$8:00$$

Draw an hour hour hand on the face of the clock below to read 8 o'clock.

Match the digital clock to the face clock.

 $7:00$

 $11:00$

         $2:00$

Digital Clocks

Write the time that is on both the digital clock and the face clock.

1.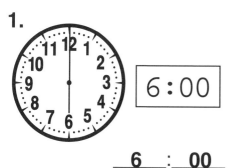

   _6_ : _00_

2.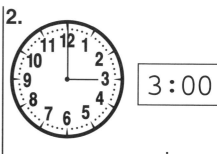

   ___ : ___

3.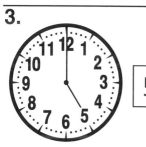

   ___ : ___

4.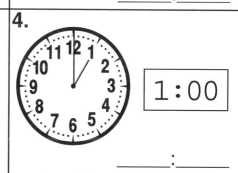

   ___ : ___

5.

   ___ : ___

6.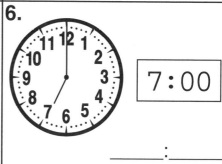

   ___ : ___

7.

   ___ : ___

8.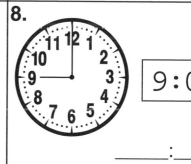

   ___ : ___

Putting Hands on the Clocks

Draw the correct hands for each clock.

1.
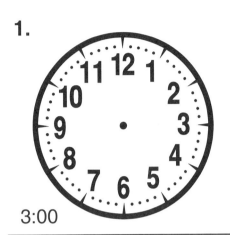
3:00

2.
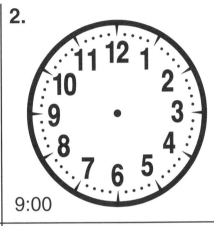
9:00

3.

7:00

4.
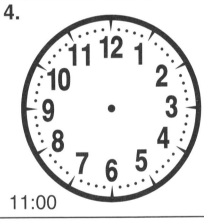
11:00

5.
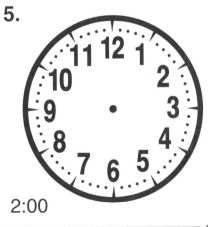
2:00

6.
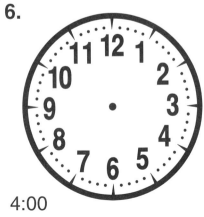
4:00

Matching Time on the Clock

Draw a line to match the time.

1.    2 o'clock

2.    9 o'clock

3.    1 o'clock

4.    6 o'clock

5.    5 o'clock

Matching Time on the Clock

Look at the time on the clock. Then, circle the correct time.

1.
1 o'clock
3 o'clock
2 o'clock

2.
10 o'clock
11 o'clock
9 o'clock

3.
2 o'clock
3 o'clock
4 o'clock

4.
8 o'clock
10 o'clock
7 o'clock

5.
5 o'clock
6 o'clock
7 o'clock

6.
7 o'clock
9 o'clock
8 o'clock

7.
3 o'clock
4 o'clock
5 o'clock

8.
12 o'clock
1 o'clock
2 o'clock

9.
12 o'clock
5 o'clock
4 o'clock

10.
6 o'clock
11 o'clock
5 o'clock

Telling Time One Hour Ahead

Practice telling time 1 hour later.
Write the first time. Then, write it 1 hour later.

1.

__7__ : __00__   1 hour later   __8__ : __00__

2.

_____:_____   1 hour later   _____:_____

3.

_____:_____   1 hour later   _____:_____

4.

_____:_____   1 hour later   _____:_____

5.

_____:_____   1 hour later   _____:_____

Telling Time One Hour Behind

Practice telling time 1 hour earlier.
Write the original time. Then, write it 1 hour earlier.

1.

   __8__ : __00__   1 hour earlier   __7__ : __00__

2.

   _____ : _____   1 hour earlier   _____ : _____

3.

   _____ : _____   1 hour earlier   _____ : _____

4.

   _____ : _____   1 hour earlier   _____ : _____

5.

   _____ : _____   1 hour earlier   _____ : _____

Read a Story about Time

Read each story. Draw the hands on the face. Write the time on the digital clock.

1. At five o'clock, Mr. Armadillo starts to cook. Yum! Ant pancakes are the best.

2. At six o'clock, Mr. Armadillo sets the table.

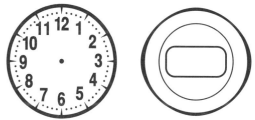

3. At seven o'clock, Mrs. Skunk reads a joke book to Baby Skunk. Baby Skunk thinks the jokes are funny.

4. At eight o'clock, Mrs. Skunk tells Baby Skunk that it is time for bed.

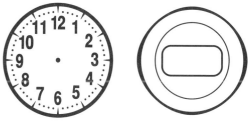

Read a Story about Time

Read each story. Draw the hands on the face. Write the time on the digital clock.

1. Mrs. Frog's babies are hungry. It is four o'clock and almost time for supper.

2. At five o'clock, the tadpoles wash up to eat their favorite food: stew and bread with jam.

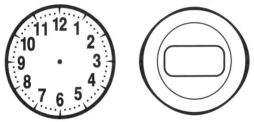

3. It is one o'clock and time for Baby Seal to learn to swim.

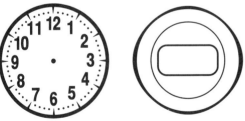

4. It is two o'clock, and Papa Seal gives Baby Seal a fish because he swam so well.

Writing the Minutes

Let's look at the big hand. The big hand is called the **minute hand**. When the minute hand goes all the way around the clock, it has been 60 minutes, or 1 hour. Look at the clock. Write the minutes on the clock.

Telling Time to the Half Hour

When the minute hand moves all of the way around the clock, that is called an hour. When it moves only halfway, that is called a **half hour**.

We have different ways to say that the time is at the half hour, or halfway around the clock.

Sometimes, we say, "It is half past nine." Sometimes, we say, "It is nine-thirty."

A half hour is written like this: 8:30.

That is because the big hand goes around the clock 30 minutes, or halfway around. The hour hand also moves halfway to the next number.

What time is it?

1. It is half past _____. It is _____:30.

2. It is half past _____. It is _____:30.

Telling Time to the Half Hour

Look at the time on the clock. Then, write the time.

1.

___4___ : ___30___

2.

_____ : _____

3.

_____ : _____

4.

_____ : _____

5.

_____ : _____

6.

_____ : _____

Writing Time to the Half Hour

Write the time that is on both the digital clock and the face clock.

1.

   _5_ : _30_

2.

   ___ : ___

3.

   ___ : ___

4.

   ___ : ___

5.

   ___ : ___

6.

   ___ : ___

7.

   ___ : ___

8.

   ___ : ___

Matching Time to the Half Hour

Choose which time is on the clock. Circle the correct time.

1.
3:30
4:30
5:30

2.
6:30
7:30
8:30

3.
11:30
12:30
1:30

4.
1:30
2:30
3:30

5.
4:30
5:30
6:30

6.
10:30
11:30
12:30

7.
7:30
8:30
9:30

8.
2:30
3:30
4:30

Putting Hands on the Clocks

Put the hands on the clocks.

1. 3:30

2. 9:30

3. 7:30

4. 11:30

5. 2:30

6. 4:30

7. 12:30

8. 1:30

Putting Hands on the Clocks

Put the hands on the clocks.

1. 8:30

2. 5:30

3. 7:30

4. 11:30

5. 1:30

6. 6:30

7. 10:30

8. 2:30

Matching Time to the Half Hour

Draw a line to match the clocks and the times.

1.    2:30

2.    9:30

3.    1:30

4.    6:30

5.    5:30

Digital Clocks to the Half Hour

This is a digital clock. It tells time with numbers. First, it tells the hour, then the minutes.

8:30

Draw hands on the face of the clock below to read 8:30.

Match the digital clock to the face clock.

1.    3:30

2.    6:30

3.    9:30

Matching Digital Clocks to Face Clocks

Draw a line to match the digital clock to the face clock.

1.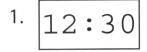

2. 1:30

3. 11:30

4. 10:30

5. 2:30

6. 9:30

7.

Counting Hours

Read each sentence. Answer the questions that follow.

Critical Thinking Skills

1. What time did the caterpillar start to spin its cocoon?

   _____:_____

   What time did the caterpillar finish?

   _____:_____

   It took the caterpillar _____ hour to spin the cocoon.

2. Mr. Beaver is building a house. What time did he start?

   _____:_____

   What time did he finish the house? _____:_____

   How many hours did it take Mr. Beaver to build his house? _____

Counting Hours

Critical Thinking Skills

Read the stories and answer the questions.

1. Mrs. Squirrel started gathering nuts at 4:00, and she finished at 6:00.
   How many hours did she spend gathering nuts?
   _____

2. Today Jimmy's pet rat had babies. She had the first one at ____:____.
   She had the last one at ____:____.
   How many hours did it take her to have all of her babies? _____

3. The space shuttle lifted off to travel around Earth at ____:____.
   It returned to Earth at ____:____.
   How many hours did it take the shuttle to go around the earth? ____

 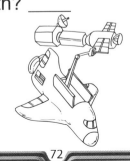 

Counting Hours and Minutes

Read the stories. The time changes in each story. Draw the hands on the face of each clock.

Critical Thinking Skills

1. It is 12:00. Heather's class goes to lunch in 30 minutes. What time will the class go to lunch?

2. Heather gets home from school at 4:30. She works on her homework for 30 minutes. What time will she be finished with her homework?

3. Heather's family plays games on Monday nights. They start to play at 8:00. They stop playing 30 minutes later. What time do they stop playing?

4. Heather goes to bed at 8:30. Then, her dad then reads a story to her for 30 minutes before she falls asleep. What time does she go to sleep?

Critical Thinking Skills

Counting Hours and Minutes

Write the times for your day. Draw the hands on each face clock. Write the time on the digital clock.

1. You are eating your breakfast.

___ : ___

2. The bus is in front of your house.

___ : ___

3. The school bell rings. You are going to class.

___ : ___

4. You can smell food. It is lunchtime.

___ : ___

5. School is out. You are headed home to play.

___ : ___

6. It is time for dinner.

___ : ___

7. The sun goes down. It is time for bed.

___ : ___

Answer Key

Page 3
1. color 5 pennies
2. color 3 pennies
3. color 7 pennies
4. color 1 penny

Page 4
color 8 pennies

Page 5
1. 3 pennies = 3¢
2. 2 pennies = 2¢
3. 1 penny = 1¢
4. 4 pennies = 4¢
5. 6 pennies = 6¢
6. 5 pennies = 5¢

Page 6
Week one: 1¢ + 1¢ = 2¢
Week two: 1¢ + 1¢ + 1¢ = 3¢
Week three: 1¢ + 1¢ + 1¢ + 1¢ = 4¢
Week four: 1¢ + 1¢ + 1¢ + 1¢ + 1¢ = 5¢
Total pennies saved: 14

Page 7
1. 3¢ 2. 5¢ 3. 8¢ 4. 7¢ 5. 4¢ 6. 2¢

Page 8
1. color 4 pennies
2. color 8 pennies
3. color 2 pennies
4. color 10 pennies
5. color 9 pennies
6. color 3 pennies

Page 9
1. 7¢ 2. 3¢ 3. 5¢ 4. 8¢

Page 10
1. 4¢ + 2¢ = 6¢
2. 1¢ + 3¢ = 4¢
3. 5¢ + 5¢ = 10¢
4. 3¢ + 2¢ = 5¢
5. 4¢ + 3¢ = 7¢
6. 4¢ + 1¢ = 5¢

Page 11
1. 4¢ − 2¢ = 2¢
2. 3¢ − 1¢ = 2¢
3. 5¢ − 5¢ = 0¢
4. 6¢ − 2¢ = 4¢
5. 4¢ − 3¢ = 1¢
6. 8¢ − 1¢ = 7¢

Page 12
A. 2¢ + 3¢ = 5¢
B. 3¢ + 2¢ = 5¢
C. 6¢ + 4¢ = 10¢
D. 4¢ + 6¢ = 10¢
E. 1¢ + 5¢ = 6¢
F. 5¢ + 1¢ = 6¢
G. 2¢ + 4¢ = 6¢
H. 4¢ + 2¢ = 6¢
I. 7¢ + 3¢ = 10¢
J. 3¢ + 7¢ = 10¢

Page 13
Answers will vary.

Page 14
25 35 50 65 70 80 90 100

Page 15
1. This is 5 cents.
2. This is 15 cents.
3. This is 20 cents.
4. This is 10 cents.
5. This is 25 cents.
6. This is 30 cents.

Page 16
1. color 2 nickels
2. color 1 nickel
3. color 4 nickels
4. color 5 nickels
5. color 6 nickels
6. color 3 nickels

Page 17
1. 15¢ 2. 10¢ 3. 20¢ 4. 30¢
5. 25¢ 6. 40¢ 7. 35¢ 8. 45¢

Page 18
Week one: 5¢ + 1¢ = 6¢
Week two: 5¢ + 1¢ + 1¢ = 7¢
Week three: 5¢ + 5¢ + 1¢ + 1¢ = 12¢
Week four: 5¢ + 5¢ + 5¢ + 1¢ + 1¢ = 17¢
You have 42¢ altogether.

Page 19
1. 5¢ 2. 7¢ 3. 8¢ 4. 11¢
5. 16¢ 6. 17¢ 7. 21¢

Page 20
1. 3 nickels = 15¢
   1 nickel and 2 pennies = 7¢
2. 3 nickels and 4 pennies = 19¢
   4 nickels = 20¢
3. 2 nickels and 2 pennies = 12¢
   6 nickels = 30¢
4. 1 nickel and 3 pennies = 8¢
   5 nickels and 3 pennies = 28¢

Page 21
1. color 1 nickel
2. color 1 nickel and 3 pennies
3. color 1 nickel and 4 pennies
4. color 2 nickels or 1 nickel and 5 pennies
5. color 1 nickel
6. color 1 nickel and 2 pennies
You have spent 44¢ total.

Page 22
1. 22¢ 2. 17¢ 3. 14¢ 4. 15¢
5. 18¢ 6. 22¢ 7. 12¢

Page 23 (Answers may vary.)
15¢ milk   20¢ burger
30¢ grapes   10¢ cupcake

Answer Key

Page 24
1. 3¢ + 1¢ = 4¢   2¢ + 4¢ = 6¢   3¢ + 5¢ = 8¢
2. 2¢ + 1¢ = 3¢   5¢ + 3¢ = 8¢   1¢ + 4¢ = 5¢
3. 2¢ + 3¢ = 5¢   4¢ + 5¢ = 9¢   3¢ + 2¢ = 5¢
4. 3¢ + 4¢ = 7¢   1¢ + 3¢ = 4¢   5¢ + 1¢ = 6¢

Page 25
30   50   70   100

Page 26
1. This is 10 cents.   2. This is 30 cents.
3. This is 60 cents.   4. This is 20 cents.
5. This is 80 cents.   6. This is 50 cents.
7. This is 40 cents.   8. This is 70 cents.

Page 27
Answers will vary.

Page 28
1. 20¢   2. 50¢   3. 70¢   4. 40¢
5. 10¢   6. 80¢   7. 60¢   8. 30¢

Page 29
1. 12¢   2. 11¢   3. 23¢   4. 15¢
5. 34¢   6. 40¢   7. 51¢

Page 30
1. color 1 dime and 1 nickel
2. color 3 dimes and 1 nickel
3. color 1 dime and 2 nickels
4. color 6 dimes and 1 nickel
5. color 2 dimes and 4 nickels
6. color 7 dimes
7. color 9 dimes and 1 nickel

Page 31
1. 40¢   2. 66¢   3. 29¢
4. 92¢   5. 54¢   6. 96¢

Page 32
1. 8¢ − 5¢ = 3¢   2. 10¢ − 2¢ = 8¢
3. 6¢ − 4¢ = 2¢   4. 12¢ − 3¢ = 9¢
5. 18¢ − 7¢ = 11¢   6. 20¢ − 1¢ = 19¢

Page 33
12¢ I bought a kite.
12¢ I bought a baseball.
5¢ I bought a jump rope.

Page 34
50   125   175

Page 35
Answers will vary.

Page 36
1. yes   2. no   3. yes   4. yes
5. yes   6. yes   7. yes   8. yes

Page 37
1. 1¢ penny   2. 5¢ nickel   3. 10¢ dime
4. 5¢ nickel   5. 10¢ dime   6. 1¢ penny

Page 38
Answers will vary.

Page 39
Answers will vary.

Page 40
42¢

Page 41
doll and scooter, no

Page 42
Answers will vary.
1. 1 dime, 2 pennies
2. 1 quarter, 1 penny
3. 1 quarter, 1 dime, 3 pennies
4. 1 quarter, 2 dimes, 4 pennies

Page 43
1. 6¢   20¢   yes   14¢
2. 17¢   27¢   yes   10¢
3. 41¢   45¢   yes   4¢

Page 44
Answers will vary.

Page 45
(clock face with numbers 1–12)

Page 46
1. 12
2. 3

Answer Key

Page 47
12   5   5 o'clock
1. It is 2 o'clock.   **2.** It is 4 o'clock.
3. It is 7 o'clock.   **4.** It is 9 o'clock.

Page 48
1. 2   **2.** 5   **3.** 6   **4.** 7   **5.** 9   **6.** 11

Page 49
1. 3 o'clock   **2.** 6 o'clock
3. 9 o'clock   **4.** 1 o'clock
5. 7 o'clock   **6.** 5 o'clock

Page 50
1.  2.
3.  4.
5.  6.

Page 51
1. 8 o'clock   **2.** 10 o'clock
3. 1 o'clock   **4.** 3 o'clock
5. 7 o'clock   **6.** 9 o'clock

Page 52
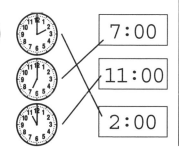

Page 53
1. 6:00   **2.** 3:00
3. 5:00   **4.** 1:00
5. 11:00   **6.** 7:00
7. 2:00   **8.** 9:00

Page 54
1.  2.
3.  4.
5.  6.

Page 55

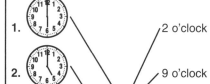

Time and Money Grades 1–3—RBP0814

Answer Key

Page 56
1. 3 o'clock
2. 10 o'clock
3. 2 o'clock
4. 8 o'clock
5. 6 o'clock
6. 7 o'clock
7. 5 o'clock
8. 1 o'clock
9. 4 o'clock
10. 11 o'clock

Page 57
7:00 1 hour later 8:00
3:00 1 hour later 4:00
11:00 1 hour later 12:00
5:00 1 hour later 6:00
1:00 1 hour later 2:00

Page 58
8:00 1 hour earlier 7:00
9:00 1 hour earlier 8:00
7:00 1 hour earlier 6:00
10:00 1 hour earlier 9:00
3:00 1 hour earlier 2:00

Page 59

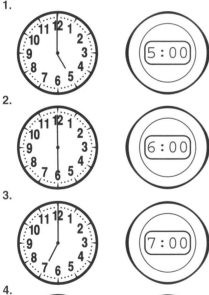

Page 60

1.
2.
3.
4.

Page 61

Answer Key

Page 62
1. It is half past 2.   It is 2:30.
2. It is half past 10.   It is 10:30.

Page 63
1. 4:30   2. 3:30   3. 5:30
4. 9:30   5. 11:30   6. 8:30

Page 64
1. 5:30   2. 2:30   3. 8:30
4. 4:30   5. 3:30   6. 6:30
7. 1:30   8. 9:30

Page 65
1. 3:30   2. 7:30   3. 12:30
4. 2:30   5. 5:30   6. 10:30
7. 9:30   8. 2:30

Page 66

Page 67

Page 68

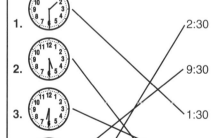

Answer Key

Page 69

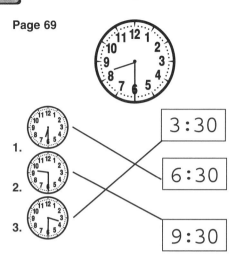

Page 70

1. 12:30
2. 1:30
3. 11:30
4. 10:30
5. 2:30
6. 9:30
7. 5:30

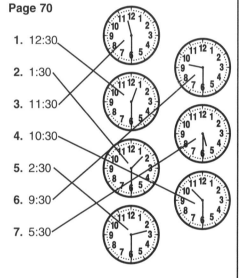

Page 71
1. 2:00 3:00 1 hour
2. 8:00 10:00 2 hours

Page 72
1. 2 hours
2. 2:00 5:00 3 hours
3. 8:00 10:00 2 hours

Page 73
1.
2.
3.
4.

Page 74
Answers will vary.

the child she has borne? Though she may forget, I will not forget you!" (Isaiah 49:15). The closest thing on this earth to our God's compassion for us is the compassion of a godly mother. Our God is a loving and nurturing God. In fact, the Bible tells us that love comes from God, and God is love (1 John 4:8). I have heard it said that no one would ever love you like your mother. I think this statement is partially true. Aside from God, no one will ever love you like your mother. Dad and Mom, you can demonstrate the nature of God to your children by loving them no matter what.

My wife is much better at showing compassion for our children than I am. I remember a time when my youngest son was pulling our puppy dog's ear. He had been repeatedly told to play nicely with the puppy. After a little time, my attention drifted elsewhere, and my son went back to making life difficult for the family pet. Finally, the dog had had enough and abruptly put a stop to the torture with the only means available to him. I couldn't blame the dog for nipping the child; he was only protecting himself. I could tell by my son's cry that he was not really hurt; he was just startled, and his feelings were damaged a little. My response to our child went something like, "See? I told you to be nice to the puppy." But whether my children incur physical or emotional pain, they look for their mother, and I am thankful my wife's response was a little different from mine. She wrapped her arms around our son and loved him up so that he would feel better.

Scripture compares the comfort of God to the comfort of a mother: "As a mother comforts her child, so will I comfort you" (Isaiah 66:13). There is nothing more comforting to a child than the embrace of his or her mother. Parents, when you show compassion to your child, whether he deserves it or not, you demonstrate a powerful attribute of God to your child.

God as an Adoptive Parent

Scripture is replete with images of God being an adoptive parent. One of the clearest Scriptures describing the adoptive fatherhood of God is found in the book of Ephesians: "For he chose us in him before the creation of the world to be holy and blameless in his sight. In love he predestined us to be adopted as his sons through Jesus Christ, in accordance with his pleasure and will—to the praise of his glorious grace, which he has freely given us in the One he loves" (Ephesians 1:4–6). God through his son Jesus makes us his own children. In fact God's Word tells us that in adoption we have been made co-heirs with Christ (Romans 8:17).

Adoption is a beautiful portrayal of the grace of God. I met a man once who had adopted a baby. He said that when the adoption finally came through he, his wife,